Oxford International Primary Geography

Workbook

M000239331

4

Terry Jennings

OXFORD
UNIVERSITY PRESS

Great Clarendon Street, Oxford, OX2 6DP, United Kingdom

Oxford University Press is a department of the University of Oxford. It furthers the University's objective of excellence in research, scholarship, and education by publishing worldwide. Oxford is a registered trade mark of Oxford University Press in the UK and in certain other countries

British Library Cataloguing in Publication Data
Data available

978-0-19-831012-9

20 19 18 17 16

Paper used in the production of this book is a natural, recyclable product made from wood grown in sustainable forests. The manufacturing process conforms to the environmental regulations of the country of origin.

Printed in Hong Kong by Sheck Wah Tong Printing Press Ltd.

Acknowledgements

The publishers would like to thank the following for permissions to use their photographs:

Cover photo: feiyuezhangjie/Shutterstock

Although we have made every effort to trace and contact all copyright holders before publication this has not been possible in all cases. If notified, the publisher will rectify any errors or omissions at the earliest opportunity.

Links to third party websites are provided by Oxford in good faith and for information only. Oxford disclaims any responsibility for the materials contained in any third party website referenced in this work.

Contents

Our environment

Human and physical features

- Walk around your local area with a responsible adult.

- Choose somewhere to stop. Use a compass to investigate what you can see to the north, east, south and west.

- Now draw and label or describe what you can see in each of the four directions.

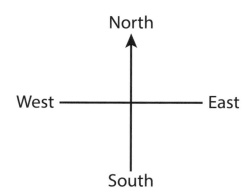

- When you are back in the classroom, look at your pictures or descriptions.

- Underline all the human features in red. These are the buildings and objects made by people.

- Underline all the physical features in green. These are natural features such as hills, rivers, lakes and forests.

- Did you see more human or more physical features? _____

Renewable and non-renewable resources

A resource is something that can be used, such as energy or materials.

Renewable resources are natural and can be replaced.

Non-renewable resources can be used once only. New ones are not formed, except over millions of years.

- Decide whether the resources named below are renewable or non-renewable. Write each one under the correct heading.

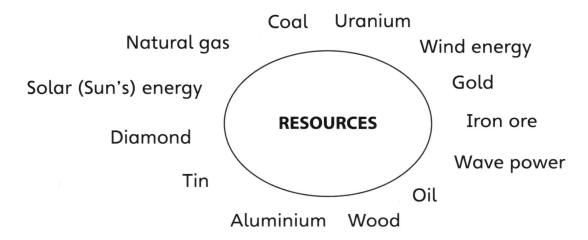

Renewable resources	Non-renewable resources

- Choose one of the non-renewable resources from your list. Suggest ways of saving it so that supplies of it will last longer.

Non-renewable resource:

_____

How it could be saved:

_____

Rubbish and litter

Litter survey

- Choose two places that you can see from the school playground or that you pass on your way to school.
- Look at (but do NOT touch) any litter in each of these two places. Fill in the details below.

	Place 1 _____	Place 2 _____
What kinds of litter are there?		
How much litter is there (one piece, two or three pieces, or more)?		
Where is the litter? (What is it in, under or behind?)		
Could it do any harm? How?		

- Is there any graffiti in either of the two places? If so, say

  where it is and what harm it is doing. _____

  _____

- Work with a friend. Discuss what you could do to improve the appearance of the two places you have studied.

A rubbish bin survey

Do people use the rubbish bins in your school grounds?

- In the large box below, draw a plan of your school grounds.

- Draw the rubbish bins on your plan.
- Now put a cross on all the areas where there is litter that is not in a bin.
- Work in groups. Each group should choose a rubbish bin.
- Weigh all the rubbish in your bin.

    The rubbish in our bin weighs _____ kilograms.

    What is the total mass of all the rubbish in all the rubbish

    bins? _____ kilograms.

- Now, wearing gloves, collect and weigh all the litter that you can see in the school grounds. The litter weighs

    _____ kilograms.

Look at your results. Is there more rubbish in the bins or out of

the bins? _____

How much more? _____ kilograms.

Improving the environment

7

Recycling rubbish

What can we recycle?

The materials for all of the things shown in the pictures below came from the Earth.

What kind of materials were they to start with? Tick the correct box. One example has been done for you.

	Made from sand	Made from oil	Made from metal ores	Made from trees
Newspapers				✓
Plastic bags				
Cardboard				
Glass jars				
Drinks cans				
Plastic bottles				
Food cans				
Glass bottles				
Magazines				

What will happen if we keep on taking these materials from

the Earth but do not recycle them? _____

_____

The three 'Rs'

There are three ways we can help to produce less rubbish and waste. They are sometimes called the three 'Rs' – reduce, recycle, reuse.

Look at the objects in the pictures. Decide whether each object can be reduced, recycled or reused. Then put a tick against its name in the correct column.

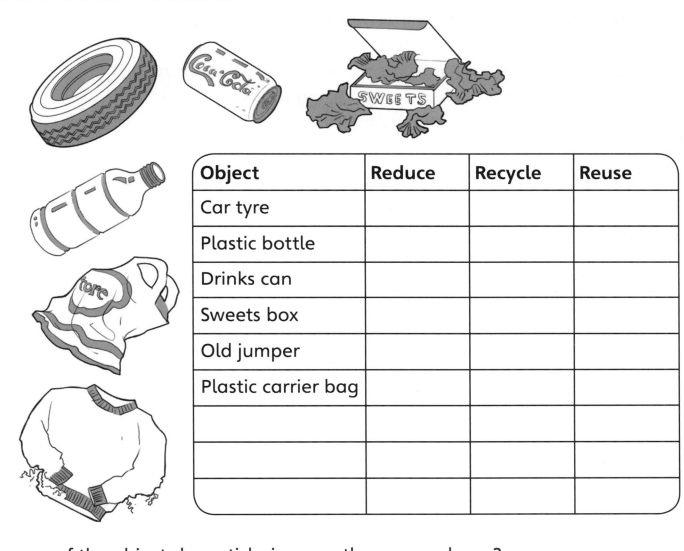

Object	Reduce	Recycle	Reuse
Car tyre			
Plastic bottle			
Drinks can			
Sweets box			
Old jumper			
Plastic carrier bag			

Do any of the objects have ticks in more than one column?

Which are they? _____

_____

Now add three more objects of your own. Which column has the

most objects in it?

_____

Oil and the environment

Relying on oil

Most of our cars and other methods of transport rely on diesel oil or petrol, which are produced from crude oil. Some of the other fuels that are produced from crude oil include liquefied petroleum gas (LPG), kerosene and jet fuel. The lubricating oil that we use to make machines run easier is also made from crude oil – and so are asphalt for surfacing roads, synthetic rubber, cosmetics, plastics, many medicines and also detergents and other cleaning products.

Think about how crude oil is obtained and how it is used. Write down all the advantages and disadvantages of using crude oil in these various ways.

Advantages	Disadvantages

How we use fossil fuels

Here is some information about how the fossil fuels – coal, oil and natural gas – were formed, and how we use them.

Put these steps in the correct order by numbering them 1 to 5.

Electricity is carried to our homes on thick cables. We use electricity to light our homes, to keep our food cool, to power the television and computer, and to do many other things.

By digging mines on land we find coal. By drilling on land or under the seabed, we find oil and natural gas and use pipelines or tankers to collect them.

About 400 million years ago, swampy forests covered much of the land. Tiny one-celled plants lived in the rivers and seas.

Coal, oil and natural gas are transported to power stations. There they are burned to produce energy, which is used to turn the generators that produce electricity.

Over millions of years, the dead remains of the plants in the swampy forests formed coal. The dead remains of the tiny one-celled plants and animals turned into oil and natural gas.

How many other uses of oil and natural gas can you think of?

_____

_____

How many other uses of electricity can you think of?

_____

_____

Energy and the environment

Global warming

In the picture, a gardener is looking after his young plants, which are growing in a greenhouse.

For each of the statements below, write TRUE if you think it is correct and FALSE if you think it is incorrect.

1 The glass in the greenhouse traps the Sun's heat by letting more in than it allows to leave.

2 All the heat the Earth gets from the Sun is trapped near to the Earth and its atmosphere.

3 The Earth has NOT become warmer over the past 100 years.

4 Some gases in the air are called 'greenhouse gases' because they act like the glass in a greenhouse.

5 We could control the amount of greenhouse gases in the air if we really tried.

6 It is the gases produced by human activities that make the Earth warm enough for us to be able to live here.

7 People burning fossil fuels – coal, oil and natural gas – have increased the amount of greenhouse gases in the air.

8 Burning forests has no effect on the amount of greenhouse gases present in the Earth's atmosphere.

9 By planting new forests, we will reduce the amount of greenhouse gases in the air.

10 If global warming continues, the ice around the North and South Poles will melt and low-lying countries such as the Maldives, the Netherlands and Bangladesh could be flooded.

Saving energy and the environment

Here are some ways of saving energy that we can use.

For each way of saving energy say whether we can do this at home (H), in the garden (G) or out shopping (S).

- ☐ Use reusable bags.
- ☐ Do not leave computers and televisions on standby.
- ☐ Collect rainwater to water plants.
- ☐ Do not fill the kettle to make just one cup of a hot drink.
- ☐ Use a compost bin.
- ☐ Put on an extra jumper in cold weather rather than turning up the heat.
- ☐ Use natural light when possible.
- ☐ Grow organic vegetables.
- ☐ Use energy-efficient light bulbs.
- ☐ Walk or cycle short distances.
- ☐ Wash only a full washing machine of clothes.
- ☐ Shower instead of taking a bath.
- ☐ Install solar panels.
- ☐ Water plants with cold washing-up water.
- ☐ Buy food that was grown or produced locally.
- ☐ Open a window instead of turning up the air conditioning.

Tick each of the things above that you are able to do.

Which of these energy saving things is most important? _____

_____

Say why. _____

_____

_____

It's a noisy world!

What a noise!

- Put your hands over your ears and close your eyes.
  Count to 10 slowly.
- Uncover your ears, open your eyes and LISTEN.
  What can you hear?
- Write all the sounds you can hear. Put a tick in the correct column for each sound. An example has been done for you.

Sound	Loud	Quiet	Continuous	Occasional	Natural	Man-made
Birds singing		✓		✓	✓	

Repeat this activity at other times and in other places, for example in the playground, in a park or garden and in your home.

Are there different sounds at different times of the day? _____

Can any of the noise be called pollution? _____

Could it be prevented? If so, how? _____

_____

It's a noisy world!

Noise is sound that is annoying or unwanted. It is possible to cut down a lot of noise by using insulation. Find out more about insulation and insulators in this activity:

- Set an alarm clock to go off in five minutes' time.

- Put your alarm clock in a small cardboard box, such as a shoe box. Close the lid and listen to the sound the clock makes when the alarm goes off.

- Now repeat the activity, but this time pack the space around the clock with newspaper before you put the lid on the box. When the alarm goes off, is it any quieter?

- Now try the activity with other packing materials and record your results below. Add two packing materials of your own.

Insulation material	Good sound insulator (makes the sound quieter)	Poor sound insulator (does not make the sound quieter)
Newspaper		
Sponge		
Polystyrene		
Cotton wool		
Straw		

- Which is the best sound insulation material that you used? _____

- Where would you put insulation in a house to keep out noise? _____

_____

Improving our environment

My favourite local area

Which place in your local area do you like best? It might be a garden, a play area, the local park, your school grounds or somewhere else.

_____

Draw your favourite place.

List three things that spoil your favourite place. _____

_____

Now draw your favourite place as you would like it to be.

Improving my local area

Choose an area near your home or school that you think could be improved.

Think of three things that could be done to improve the area.

Name of area: _____

Three ways to improve it:

1 _____

2 _____

3 _____

Draw a picture of the area showing your improvements.

Improving my local area

Early villages

Sorting buildings

We can sort buildings into different sets. Each set describes what the building is used for.

1 Homes – where people live.

2 Transport – where people go to travel.

3 Leisure – where people have fun.

4 Shops – where people buy goods.

5 Offices, factories – where people go to work.

• Write a list of the different kinds of buildings in your area. Put each building in the correct set.

• Which set has the fewest buildings in it?

• Do any buildings appear in more than one set?

• Which set would you add to in order to improve your local area?

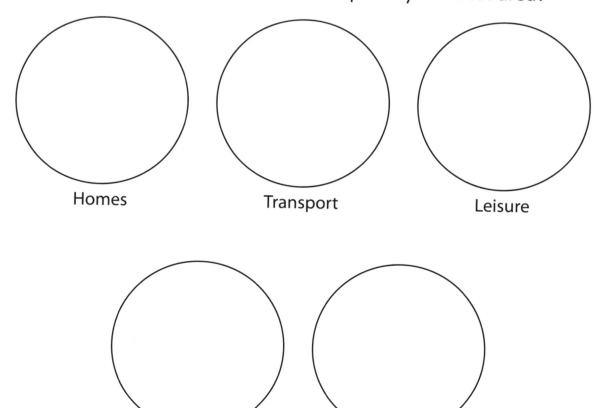

Homes   Transport   Leisure

Shops   Offices/factories

A new settlement

A settlement is a place where people live. Villages, towns and cities are all settlements.

Imagine you are living about a thousand years ago. You and your family and friends are looking for a site for a new settlement. Below is a list of things you might want to find in the site. For each of these things, explain why you think it is necessary.

Feature	Why it is necessary
Clean water	
Dry site	
Safe from attack	
Shelter	
Building materials	
Fertile soil	
Supply of fuel	
Aspect (which way the site faces)	
Communications	

Which of these features do you think is most important? Say why. _____

_____

Which of these features do you think is least important? Say why. _____

_____

Balad Sayt village

The village of Balad Sayt

Here is a map showing the position of the village of Balad Sayt in Oman.

Answer these questions:

a What is the capital city of Oman?

_____

b In which direction is Balad Sayt from the capital of Oman?

_____

c Roughly how far is it from Balad Sayt to the capital?

_____

d Which sea lies to the east of Oman?

_____

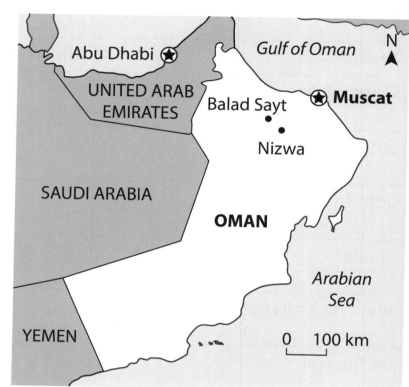

e Which gulf lies to the north of Muscat? _____

f What is the name of the city to the south-east of Balad Sayt? _____

g If you travelled west from Balad Sayt, which country would you

eventually reach? _____

_____

h Name two other countries that border Oman. _____

_____

Oman fact file

The village of Balad Sayt is in Oman. Find out all you can about Oman.
Fill in the fact file.

Continent	
Population	
Capital city	
Currency (money)	

Flag

Languages spoken	
Other big cities	
Neighbouring countries	

Famous sights and interesting facts

How a village grows

How a village grows

Many villages grew into towns. They became meeting places, usually for buying and selling goods. The most common kind of town was a market town used by farmers.

Can you find the names of four small market towns in your country? They will have a market where fruit, vegetables and animals, produced on farms, are sold for food.

_____   _____

_____   _____

Coastal villages often began with people catching and selling fish. This was the main function of coastal villages.

As these villages grew, their functions often changed or they developed more than one function.

What are the functions of these coastal towns? Choose the correct label from the word box.

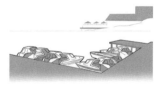

_____   _____

_____   _____

| seaside resort | harbour | ferry port | oil terminal |

Village changes

Here are two pictures of the same village. The top picture shows the village 60 years ago. The bottom picture shows the village as it is today.

- Colour in any changes that have taken place.

- Do you think the changes are good or bad for the village? Say why. _____

_____

_____

India

India

Use an atlas to help you with this activity.

1 What is the name of this continent? _____
Colour all the oceans and seas blue.

2 Find India on the map and label the following features:

    a New Delhi, the capital city of India

    b the cities of Mumbai, Kolkata and Bangalore

    c the River Ganges

    d Sri Lanka, an island country near the southern tip of India

    e the Himalayas (this mountain range is only partly in India).

3 Draw lines to show the Equator, the Tropic of Cancer and the Tropic of Capricorn.

4 Colour India yellow.

5 Colour the other land areas green.

Comparing countries

- Compare India, your country and another country you have been to or are interested in. Use reference books and the Internet to help you complete the table. One item has been done for you.

	My country _____	India	Another country _____
Continent			
Language(s)			
Climate			
Currency (money)		Rupee	
Food			
Transport			
Work			
Leisure			
School			

Country life in India

Amina's day

Amina is 10 years old. She lives in a village in northern India.

This clock shows how she spends her day.

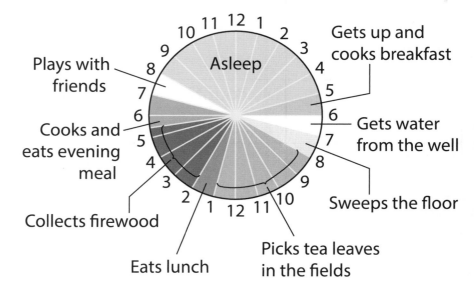

- How much time does she spend on each activity?

Activity	Amount of time
Sleeping	
Working in the fields	
Collecting firewood	
Preparing meals	
Fetching water	
Playing with friends	

- Repeat the activity above for yourself, showing the amount of time you spend on the things you do each day.

How is your day similar to Amina's? How is it different? _____

_____

_____

_____

Daily diet

Ambika is 10 years old and she lives in a village in southern India. This is the food eaten by Ambika in a typical day:

Breakfast: A bowl of cold runny 'porridge' called sorghum. Water from the village well to drink.

Lunch: Rice with vegetables. Water to drink.

Supper: Boiled millet and samba (curried vegetables and lentils). Tea to drink.

The rice, sorghum, millet, vegetables and lentils are grown near the village. The tea is bought from the village shop but is grown in other, cooler parts of India. The drinking water comes from the village well – a five-minute walk away.

- Write a list of all the food and drink you consume in a typical school day. Where is it grown or produced? (Read the labels on the cartons, tins and jars for this information.)

Food or drink	Where it was grown or produced

1 Do you eat more varieties of food or fewer varieties of food than Ambika eats? _____

2 Do your foods come from nearer to your home or further away than Ambika's foods do? _____

Comparing rainfall

Different parts of the world have very different weather. The word 'weather' describes the daily changes in sunshine, cloud, wind and rainfall at any one place. The average weather of a place over the course of one year is called its climate.

Rainfall (mm)	Jan	Feb	Mar	Apr	May	Jun	Jul	Aug	Sep	Oct	Nov	Dec
Kolkata	10	31	36	43	140	297	325	328	252	114	20	5
Paris	56	46	54	47	63	58	54	52	54	56	56	56

The data above shows the average amount of rain that falls in Kolkata, India, each month during the year. Complete the graph, using the data above. Plot the rainfall for Paris, France, in a different colour.

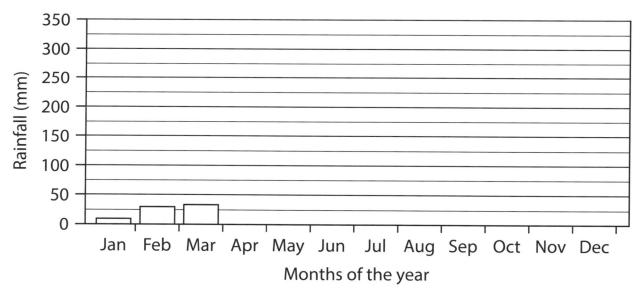

1 How much rain falls in Kolkata in an average year? _____

2 How much rain falls in Paris in an average year? _____

3 Which are the wettest months in Kolkata? _____

4 Which are the driest? _____

5 Which are the wettest months in Paris? _____

6 Which are the driest? _____

Comparing temperatures

As well as having different rainfall, different countries have different temperatures.

Temp °C	Jan	Feb	Mar	Apr	May	Jun	Jul	Aug	Sep	Oct	Nov	Dec
Kolkata	27	29	34	36	36	33	32	32	32	32	29	26
Paris	6	8	11	14	18	22	24	24	21	16	10	7

Use the data above to complete the graph showing the average maximum temperatures each month in Kolkata, India.

Plot the temperatures for Paris, France, in a different colour.

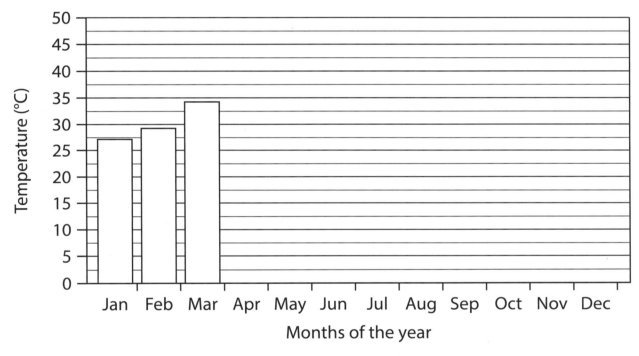

Now answer these questions:

1 Which are the hottest months in Kolkata? _____

2 Which are the hottest months in Paris? _____

3 What is the difference between the highest temperature in

   Kolkata and the highest temperature in Paris? _____

4 Which are the coldest months in Kolkata? _____

5 Which are the coldest months in Paris? _____

Life in an Indian fishing village

Village and city life in India

Here are some statements about India. For each one say whether you think it is about a village **(V)** or a city **(C)**.

a It is quiet and peaceful with clean fresh air. ☐

b There are lots of jobs all the year round and the wages are usually good. ☐

c Work is seasonal, such as farming during the rainy season, and wages are very low. ☐

d There are lots of rice fields with surrounding green countryside. ☐

e It is busy with lots of buildings, noise and traffic. The air is not clean. ☐

f Most of the homes have water and electricity. ☐

g Transport is limited to buses and autorickshaws, unless you have your own car or bicycle. ☐

h Many of the buildings are made of brick with thatched roofs. ☐

i There are lots of shops, supermarkets and street markets. ☐

j There are some very tall buildings as well as some smaller, old buildings. ☐

k Some of the homes have electricity. ☐

l Water is collected from a well by the women. ☐

m All kinds of transport are nearby. ☐

n Some women go out to work and also do the chores around the home. Some women stay at home. ☐

o There is a primary school for local children to attend. ☐

p There are few shops but some market stalls by the roadside where the locals sell their produce. ☐

q There are many primary schools, although not all children can afford to go to school. ☐

Getting to India

This is a map showing part of the world.

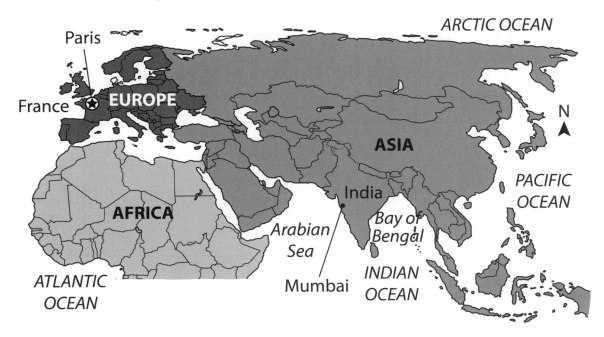

Answer these questions:

1 In which continent is France? _____

2 In which continent is India? _____

3 In which direction would you be flying if you went from Paris

   to Mumbai? _____

4 Use an atlas to find out which countries you might fly over if going from

   Paris to Mumbai. _____

5 Name five countries that border India. Use an atlas to help you. _____

   _____

6 What is the name of the island country off the southern tip

   of India? _____

Life in an Indian city

City life and country life in India

Compare life in a city in India, such as Mumbai or Kolkata, with life in an Indian village such as Parsoiya.

	an Indian city	an Indian village
Climate		
Family life		
Houses and homes		
Schools		
Jobs		
Water supply and electricity		
Journeys and transport		
Changes in the area		

Repeat the activity, comparing life in a city in your country with life in a village in your country.

What things are similar in India and your country? _____

_____

_____

What things are different? _____

_____

India fact file

What have you learned about India?

Fill in the fact file.

Continent	
Population	
Capital city	
Currency (money)	

Flag

Languages spoken

Other big cities

Neighbouring countries

Famous sights and interesting facts

Leisure time

Work, leisure and recreation

Work is something you have to do, including school work or jobs for your parents or carers. Leisure is a time that is free from work, when you can do what you like.

Recreation is a game, hobby or enjoyable pastime you do in your spare time.

- Keep a diary, every day for a week, of the length of time you spend doing work and being at leisure or recreation.

- Dates: _____

Day	At school			At home		
	Work	**Leisure**	**Recreation**	**Work**	**Leisure**	**Recreation**
Monday						
Tuesday						
Wednesday						
Thursday						
Friday						
Saturday						
Sunday						

- Choose a way to present what you have recorded in your diary. You could use a spreadsheet or some graphing software.

- How many hours of the week do you spend doing school work at school? _____

- How many hours do you spend doing school work at home? _____

- How many hours do you spend at leisure? _____

- How many hours do you spend on recreational activities? _____

Where do we go for our leisure activities?

Work with a group of friends.

Discuss what you all like to do when you have time to yourself.

Where do you go for these leisure activities?

Write a list of each activity and the place where you go to do it.
This has been started for you.

Activity	Where we go
Watching television	*At home*

For which of the activities do you need a special area of land? _____

_____

For which of the activities do you need a special room or building? _____

_____

Leisure and land use

Leisure activities

Think about each of the leisure activities below. Which of them can you do best at home (**H**), which in a park or playground (**P**) and which need a special place (**S**)? Write the correct letter next to each one.

badminton ☐   swimming ☐   reading ☐   golf ☐

cards ☐   bingo ☐   table tennis ☐   roller blading ☐

judo ☐   gymnastics ☐   bowls ☐   tennis ☐

Imagine you have been given an old factory to convert into a leisure centre.

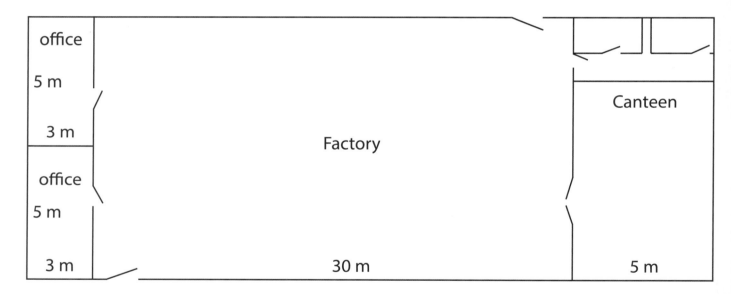

On the plan, show what activities would take place in the different parts of the building.

What extra equipment would you need? _____

_____

_____

_____

Leisure and land use

Look at the leisure activities in the pictures below.

Golf   Skiing   Football   Tennis

Table tennis   Roller blading   Ten-pin bowling

Which of these activities usually need a special area of land or

a special building? _____

_____

Which of them need a large area of land? _____

Which of them needs the smallest amount of land? _____

Where would you usually expect to see people skiing? _____

Would you expect to find a golf course near the middle of a city? _____

Say why. _____

Earthquakes and volcanoes

Earthquakes and volcanoes

Find information about earthquakes and volcanic eruptions in the following ways. Watch the television news reports. Listen to radio news reports. Study the newspapers and the Internet.

Record any earthquakes and volcanic eruptions. Fill in a box for each one saying what happened.

From each box, draw a line to the correct country on the map of the world to show where it happened.

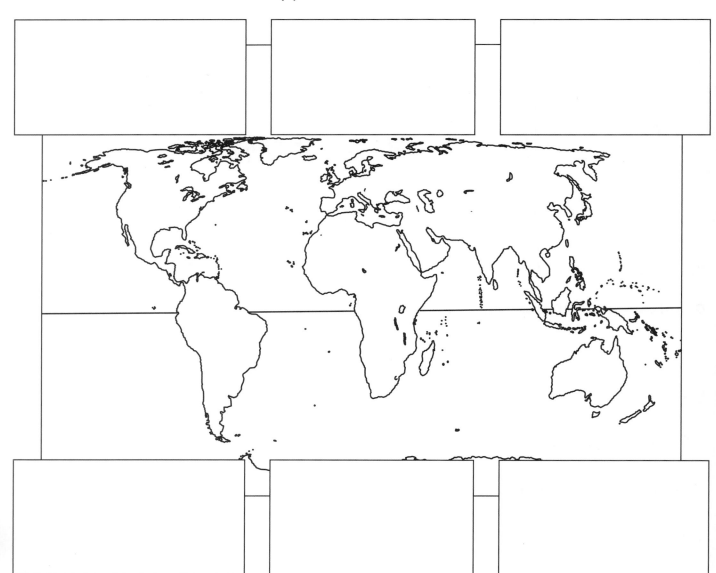

A model of the Earth's plates

- On a small map of the world, mark the edges of the Earth's plates, as shown on the map below.

- Stick your map onto a sheet of card.

- Carefully cut out the plates.

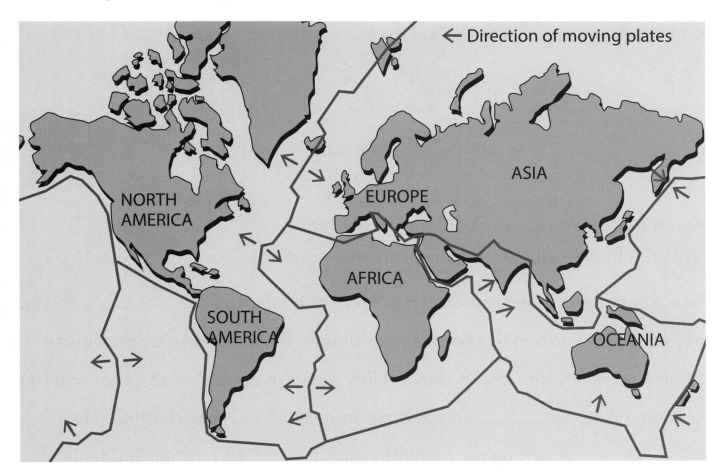

← Direction of moving plates

NORTH AMERICA

EUROPE

ASIA

AFRICA

SOUTH AMERICA

OCEANIA

Experiment with your plates to answer these questions:

1 a What will happen to the Pacific Ocean if the Atlantic Ocean continues to get wider? You can use an atlas if you need to.

_____

_____

b Which continents will move? _____

2 What will happen if Africa moves closer to Europe? _____

_____

Earthquakes

What is an earthquake?

How much do you know about earthquakes?
Fill in the blanks using the words from the word box.

An earthquake is caused when two or more of the Earth's _____ bump

into each other. When the Earth's plates meet, they usually slide past each

other. Sometimes though, as they collide, they rub against each other and this

rubbing, or _____, stops them from sliding over each other. The

_____ builds up as the plates push against each other. Suddenly they

slip past each other. This releases a lot of _____ and the land above

_____ violently. This energy is released as waves, called seismic

waves. It is these seismic waves that cause so much _____ in an

earthquake. The place where the earthquake actually happens is called the

_____ of the earthquake. The place where the earthquake happens

on the surface of the ground is called the _____.

| damage | plates | epicentre | pressure |
| friction | energy | focus | shakes |

Earthquake emergency!

Imagine you live in an area where a major earthquake has just occurred.

More aftershocks are likely.

Your apartment is badly damaged so you cannot stay there.

Although it is warm in the daytime, it is very cold at night.

You need to plan how to stay alive and keep safe for the next five days.

Look at this list of items.

torch	emergency food	bottled water
warm clothing	tin opener	first aid kit and first aid book
portable radio	blankets	money and credit cards
mobile phone	photographs	pens, pencils and notebook
medicines	camping stove	jewellery and watches
passport	play station	gloves, strong boots or shoes
laptop computer	television set	

Pick eight of the items that you would choose to take with you.

1 _____   2 _____

3 _____   4 _____

5 _____   6 _____

7 _____   8 _____

The most important item in my emergency kit is _____

This is because _____

_____

Volcanoes

The parts of a volcano

Use the words in the word box to label the parts of the volcano.

cone	crater	lava
magma	vent	dust, ash and rock

What is the difference between lava and magma? _____

_____

What is the name of the nearest volcano to your home or

school? _____

Which country is it in? _____

How far is it from where you live? _____

A volcano map quiz

Here is a map quiz. Use an atlas, reference books and the Internet to help you answer these questions about the map opposite.

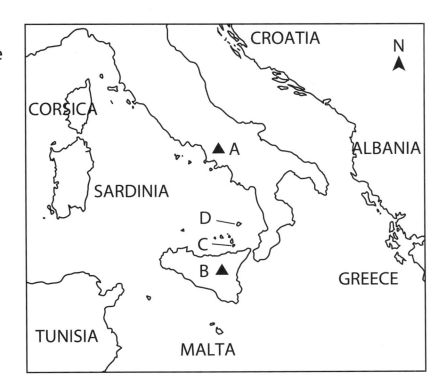

a The map shows part of a sea that is almost completely surrounded by land. What is the name of this sea?

_____

b What is the name of the country that is shaped rather like a boot?

_____

c What is the name of the volcano labelled A? _____

d What is the name of the island labelled B? _____

e What is the name of the volcano on the island labelled B?

_____

f What is the name of the extinct volcano on the small island labelled C?

_____

g There is another volcano on the tiny island labelled D. What is its name?

_____

h Of the four volcanoes marked on this map, which is the tallest?

_____

Map of the World

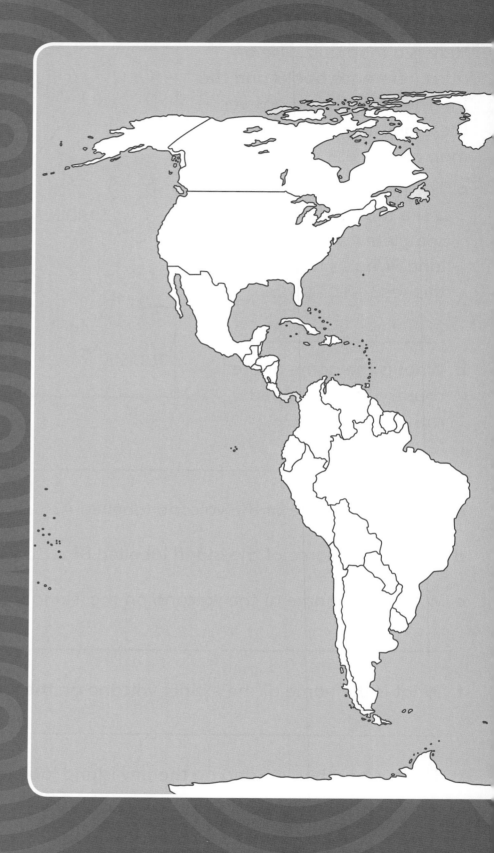

Glossary

Acid rain Rainwater that contains acids formed from harmful gases that can kill plants and animals and damage buildings.

Active volcano A volcano that still erupts.

Atmosphere The thick layer of air that surrounds the Earth.

Compost Kitchen and garden waste used to fertilise and improve the soil.

Continent One of the seven huge pieces of land in the world.

Core The centre of the Earth made of the metals iron and nickel.

Crater The cup-shaped hollow around the opening of a volcano.

Crude oil Oil formed under the ground that has not been purified.

Crust The Earth's outer layer of rock on which we live.

Dormant volcano A volcano that is 'resting' or 'sleeping', and which has not erupted in recent years.

Drought A very long period without rain.

Earthquake A movement or shaking of the Earth's crust, often caused when the Earth's plates move against each other.

Energy The power and ability something has to do work.

Environment Your surroundings.

Erupt A volcano is said to erupt when lava, dust, ashes and other volcanic materials are forced out of it.

Export Things made or grown in one country and then sold to people in another country.

Extinct volcano A term used to describe a volcano that has not erupted for at least 10 000 years.

Famine A severe shortage of food.

Fault A large crack or break in a series of rocks. The rocks on one or both sides of the fault may slip up or down.

Fertile If a soil is fertile, it is able to produce good crops.

Fertiliser A substance put on the soil to make plants grow better.

Fossil fuel A fuel such as coal, oil or natural gas that was formed from living things a very long time ago.

Global warming The build-up of carbon dioxide and other gases in the atmosphere that trap the Sun's heat, so causing the Earth's temperature to rise and its climate to change.

Irrigation The taking of water from rivers, lakes, wells or reservoirs to the land so that crops can grow well.

Jute A material obtained from the bark of certain plants that is used for making sacks, mats and ropes.

Landfill site A hole or pit in the ground used for getting rid of waste. Where rubbish is buried.

Lava The molten rock that comes out of a volcano.

Leisure time Time when you can do what you like.

Litter Rubbish that is not put in a rubbish bin.

Magma The hot, runny rock found under the surface of the Earth. If it escapes onto the surface it is then called lava.

Mantle The layer of the Earth, immediately beneath its crust. Some of the mantle is made of solid rock, while some of it is molten rock.

Monsoon A strong wind in or around the Indian Ocean that brings heavy rain in summer.

Natural resources Materials and energy that we get from the air, the Sun, animals, trees and other plants, rocks, oceans, seas and the soil.

Noise Unwanted or unpleasant sound.

Nomad Someone who moves from place to place instead of living and working in the same area.

Non-renewable resources Natural resources that cannot be replaced after they have been used.

Ore A rock that contains metal.

Plate	One of the 19 or 20 sections of rock that make up the Earth's crust.	**Rock**	One of the parts of the solid surface of the Earth.
Pollute	To make something dirty.	**Season**	A time of the year when you can expect a certain pattern of weather.
Pollution	When substances such as air, water or the soil are spoiled or made dirty by people.	**Shanty town**	An area of a city where people have built their own houses from waste materials.
Population	All the people who live in a particular place.	**Temporary**	Lasting, or meant to last, for only a short time.
Port	A place on the coast where ships can dock and transfer people and goods to and from the land.	**Terrace**	A raised level space, like a stair.
		Textile	Fabric or cloth.
Recycle	To treat waste material so that it can be used again.	**Trade**	The buying and selling of goods.
Refugee	A person who has fled from danger or a problem such as famine or war.	**Tributary**	A small river or stream that flows into a larger river.
Renewable resources	Materials or energy from sources that are constant and natural, like plants, animals, the Sun, wind and waves.	**Volcano**	A hole or tear at a weak spot in the Earth's crust from which gases and hot, molten rock flow.
		Wadi	A valley in a hot desert area which is usually dry, but may have a river or stream in it after heavy rain.